YOU WILL PASS ALGEBRA

You Will Pass Algebra

Poetry Affirmations for Math Students

Walter the Educator™

SKB

Silent King Books a WhichHead Imprint

Copyright © 2023 by Walter the Educator™

All rights reserved. No part of this book may be reproduced in any manner whatsoever without written permission except in the case of brief quotations embodied in critical articles and reviews.

First Printing, 2023

Disclaimer
This book is a literary work; poems are not about specific persons, locations, situations, and/or circumstances unless mentioned in a historical context. This book is for entertainment and informational purposes only. The author and publisher offer this information without warranties expressed or implied. No matter the grounds, neither the author nor the publisher will be accountable for any losses, injuries, or other damages caused by the reader's use of this book. The use of this book acknowledges an understanding and acceptance of this disclaimer.

dedicated to all the math lovers
across the world

CONTENTS

Dedication v

Why I Created This Book? 1

One - Witness The Wonders 2

Two - Emerge Victorious 4

Three - Conquer Algebra 6

Four - Success In Algebra 8

Five - Path You Embark 10

Six - Never Let Go 12

Seven - You Shall Pass 14

Eight - Endless Possibilities 16

Nine - Unlock The Door 18

Ten - Brilliance And Class 20

Eleven - Forever Free 22

Twelve - Leaving No Dream Behind 24

Thirteen - Conquer And Embrace	. . . . .	26
Fourteen - Gather Your Courage	. . . . .	28
Fifteen - Believe In Yourself	. . . .	30
Sixteen - Your Victory Song	. . . .	32
Seventeen - Confidence And Grace	. . . .	34
Eighteen - Essence Of Algebra	. . . .	36
Nineteen - Future That's Blessed	. . . . .	38
Twenty - Find Your Worth	. . . .	40
Twenty-One - Potential Soar	. . . .	42
Twenty-Two - Greatness Will Be	. . . . .	44
Twenty-Three - Never Give In	. . . . . .	46
Twenty-Four - Don't Let Doubts Persist	. . .	48
Twenty-Five - Journey Of Discovery	. . . .	50
Twenty-Six - Fear Not The X's Or The Y's	. .	52
Twenty-Seven - Bright And True	. . . .	54
Twenty-Eight - Passion And Drive	. . . . .	56
Twenty-Nine - Stepping Stone	. . . . . .	58
Thirty - You Will Rise	. . . .	60
Thirty-One - Doubts Will Surely Break	. . .	62
Thirty-Two - Dear Student	. . . .	64

Thirty-Three - Algebra Divine 66

Thirty-Four - Symbols And Signs 68

Thirty-Five - Your Faithful Friend 70

About The Author 72

WHY I CREATED THIS BOOK?

Creating a poetry book to motivate a student to pass the subject of Algebra can be an innovative and effective approach. Poetry has the power to engage emotions and make learning more enjoyable. By using poetic language and metaphors, this book can simplify complex algebraic concepts, making them more accessible and relatable. It can also provide encouragement and inspiration to persevere through the challenges of learning algebra. This book can serve as a creative tool to instill confidence, boost motivation, and help students develop a positive attitude towards the subject.

ONE

WITNESS THE WONDERS

In the realm of numbers, where logic plays its part,
There lies a subject, feared by many, called Algebra, an art.
Its equations and formulas may seem like a maze,
But fear not, dear student, for I'll guide you through its ways.

Algebra, a puzzle waiting to be solved,
A language of symbols, where mysteries are involved.
Through variables and constants, we'll find the unknown,
With determination and effort, your skills will be honed.

Let's dive into the depths of this numerical ocean,
Where patterns and patterns form a beautiful potion.

Solve for x, unravel the mystery,
And witness the beauty of this mathematical history.
With each step forward, you'll gain confidence anew,
Discovering connections, as the subject unveils its view.
The satisfaction of solving a complex equation,
Will fill your heart with joy, beyond explanation.
Algebra, the gateway to a world of knowledge,
Opens doors to opportunities, from college to college.
Embrace its challenges, for they'll make you grow,
And the seeds of success, in your mind, they'll sow.
So, dear student, let not fear hold you back,
For Algebra's rewards are worth the knack.
With perseverance and dedication, you shall pass,
And conquer this subject, like a true mathematical ace.
Believe in yourself, and let the numbers be your guide,
For Algebra's secrets, with you, shall reside.
Unlock the power within, and set your mind free,
And witness the wonders of Algebra, for all to see.

TWO

EMERGE VICTORIOUS

In the realm of numbers and equations,
Lies the captivating world of Algebra.
A subject that may seem daunting at first,
But fear not, for it holds treasures untold.

With its unknown variables and intricate patterns,
Algebra beckons you to unravel its mysteries.
Through the tangled web of equations,
Lies the path to knowledge and understanding.

Embrace the challenge, dear student,
For in Algebra lies the power to unlock
The secrets of the universe and beyond.
Let the numbers be your guide, your compass.

With every equation you solve,
A sense of satisfaction will fill your soul.
For in conquering Algebra, you conquer yourself,
Unleashing your hidden potential and strength.

Believe in yourself, for you are capable,
Of mastering this beautiful mathematical art.
Let Algebra be your stepping stone,
To a brighter future, to dreams realized.

For it is through Algebra's gates,
That you'll find the key to countless opportunities.
From science to finance, from engineering to art,
Algebra paves the way for personal growth.

So, fear not the challenges that lie ahead,
For with dedication and perseverance,
You shall conquer Algebra's realm,
And emerge victorious, forever transformed.

THREE

CONQUER ALGEBRA

In the realm of numbers, where patterns reside,
There lies a subject, often feared, set aside.
Algebra, they call it, with its mysterious ways,
A puzzle to unravel, in the midst of chaotic arrays.

Oh, student of numbers, do not despair,
For Algebra holds secrets beyond compare.
It is a language, a key to unlock,
The wonders of the universe, within your flock.

Equations and variables, they may seem daunting,
But fear not, for in them lies a rewarding haunting.
For when you conquer the x's and y's,
A world of possibilities before you lies.

Algebra, dear student, is a powerful tool,
To solve real-life problems, it makes you cool.

From calculating distances to predicting trends,
It empowers you, your knowledge transcends.
    So, embrace the challenge, let no fear bind,
For Algebra is the pathway to the mind.
Discover its beauty, unravel its grace,
And watch as your potential finds its place.
    Believe in yourself, with every equation you solve,
Let Algebra's magic within you evolve.
For when you pass this subject, oh so grand,
You'll hold the key to success in your hands.
    Algebra, dear student, is a journey worthwhile,
So persevere, my friend, go the extra mile.
Unlock the mysteries, unlock your fate,
And conquer Algebra, before it's too late.

FOUR

SUCCESS IN ALGEBRA

In the realm of numbers, where patterns reside,
Lies a subject that often fills hearts with fright.
Algebra, they call it, a puzzle to solve,
A challenge that can truly evolve.

Fear not, dear student, for in this maze,
Lies the key to unlock hidden ways.
Algebra, my friend, is not just a chore,
But a language of logic, a tool to explore.

With every equation, a story unfolds,
A dance of variables, mysteries untold.
Each problem you solve, a victory won,
A step closer to the brilliance you've become.

For Algebra reveals the beauty within,
The symmetry of numbers, the logic therein.

It sharpens your mind, expands your sight,
Unleashing your potential, shining so bright.

So, embrace the challenges, face them with glee,
For Algebra's rewards are yours to see.
Let numbers be your allies, equations your guide,
As you journey through Algebra, side by side.

Believe in yourself, and never give in,
With determination, you're destined to win.
Passing Algebra, a triumph well earned,
A testament to the lessons you've learned.

So, dear student, don't falter or despair,
For Algebra's conquest is yours to bear.
With perseverance and a heart full of zeal,
Success in Algebra, you shall surely reveal.

FIVE

PATH YOU EMBARK

In the realm of numbers, where mysteries lie,
A subject awaits, with questions to untie.
Algebra, the language of the unseen,
Unlocks the secrets of the universe, so keen.

    Embrace the challenge, the equations unfold,
Solve the puzzles, let your brilliance take hold.
From x to y, and variables unknown,
Algebra leads you to a world of your own.

    With every equation, a problem you'll solve,
Mastering the language, evolving and evolve.
Patterns and logic, they dance in your mind,
As Algebra reveals the treasures you'll find.

    Beyond the classroom, opportunities arise,
Algebra opens doors, reaching for the skies.

In science, engineering, or finance's embrace,
Algebra paves the way, leaving a lasting trace.

Believe in yourself, let your potential ignite,
For Algebra is a journey, a beacon of light.
Persevere through the challenges, never give in,
And watch as victory turns into a grin.

So fear not the numbers, embrace them with glee,
For Algebra is a friend, a tool that sets you free.
With dedication and effort, you'll conquer each test,
And find a world of possibilities, where dreams manifest.

So let Algebra guide you, on this path you embark,
And witness the beauty, as it ignites a spark.
For within the realm of numbers, magic takes flight,
And as you pass Algebra, your future shines bright.

SIX

NEVER LET GO

In the realm of numbers, where logic weaves its spell,
Lies the subject of Algebra, where secrets dwell.
Fear not, dear student, for I bring you a tale,
Of triumph and glory, where equations prevail.

    Algebra, a puzzle, a riddle to be solved,
A challenge to conquer, a problem to resolve.
Through variables and constants, an unknown awaits,
A world of possibilities, where knowledge creates.

    Unlock the gates of Algebra, and you shall see,
A universe of patterns, waiting to be set free.
With every equation you solve, a door swings wide,
Revealing the beauty that lies deep inside.

    From simple equations to complex domains,
Algebra empowers, it expands our brains.

It builds the foundation for all that lies ahead,
In science, in technology, in life we tread.

So fear not, young student, embrace the unknown,
For Algebra is the key to dreams yet unblown.
Discover the joy of solving each equation,
And witness the power of mathematical creation.

With Algebra as your ally, the world is at your feet,
Opportunities abound, success is yours to meet.
So persevere, believe, and never let go,
For Algebra's rewards are yours to bestow.

SEVEN

YOU SHALL PASS

In the realm of numbers, a journey begins,
Where algebra's secrets unlock and transcend.
Oh student, embrace this world so grand,
For within its depths lies wisdom's hand.

Fear not the variables that dance and sway,
For they hold the power to pave your way.
Equations and formulas may seem like foes,
But conquer them, and knowledge will expose.

Algebra, dear student, is not just a test,
But a key to success, your very best.
It opens doors to realms unknown,
Where opportunities blossom and dreams are sown.

From physics to finance, and art to design,
Algebra's magic will forever shine.
It equips you with skills, both sharp and keen,
To unravel life's puzzles, unseen.

So, persevere, dear student, through every trial,
For Algebra's challenges will shape your style.
Believe in yourself, for you hold the key,
To unlock your potential, and set yourself free.

In the realm of numbers, let your spirit rise,
As you conquer Algebra's infinite skies.
With determination and passion, you'll find,
The beauty and power that lie behind.

So fear not, dear student, for you shall pass,
And witness the wonders of Algebra's class.
Embrace the challenges, let your dreams unfurl,
For Algebra's triumph will change your world.

EIGHT

ENDLESS POSSIBILITIES

In the realm of numbers and equations,
Where Algebra reigns with its calculations,
A student stands, seeking inspiration,
To conquer this subject, with dedication.

Algebra, a puzzle of symbols and signs,
A gateway to knowledge, where wisdom aligns,
With patience and perseverance, you shall find,
The power to unlock the secrets of the mind.

Fear not the unknown, embrace it instead,
For Algebra's challenges lie just ahead,
In every equation, a lesson to be learned,
A path to success, yet to be discerned.

Through variables and formulas, you'll explore,
The depths of logic, like never before,

Solving equations, like unraveling a mystery,
Unleashing the potential within, with great victory.

For Algebra's magic lies not in its rules,
But in the growth and understanding it fuels,
With every step forward, you'll gain insight,
A seed of knowledge, blossoming bright.

So, dear student, with passion and drive,
Embrace Algebra's beauty, let it revive,
Your dreams and ambitions, set them free,
For Algebra holds the key, believe in thee.

In this realm of numbers, you'll find your way,
To conquer the challenges, day by day,
And as you pass this subject, you'll see,
The endless possibilities, that Algebra can be.

NINE

UNLOCK THE DOOR

In the realm of numbers, where mysteries reside,
There lies a subject, where many confide.
Algebra, they call it, a puzzle to solve,
A treasure trove of knowledge, waiting to evolve.

    Embrace the challenge, dear student, with might,
For Algebra holds the key to a future so bright.
It expands your mind, unlocks the unknown,
A gateway to success, you'll proudly have grown.

    Through equations and variables, you'll find your way,
Solving problems, step by step, day by day.
Believe in yourself, with perseverance, you'll see,
Algebra's power, setting your spirit free.

    In science and engineering, its relevance is clear,
In finance and economics, it's always near.

From architecture to computer science's grace,
Algebra's foundation, in every field we embrace.

Dedicate your effort, with a positive mind,
Algebra's challenges, you'll surely unwind.
Let it guide you, like a friend by your side,
Witness the magic, as numbers collide.

So fear not, dear student, let Algebra be,
The catalyst for dreams, setting your mind free.
Unlock the door to knowledge, let it be your guide,
And witness the possibilities that within Algebra reside.

TEN

BRILLIANCE AND CLASS

In the realm of numbers, where mysteries unfold,
Lies a subject called Algebra, both complex and bold.
Fear not, dear student, for I shall reveal,
The secrets that Algebra has in store, concealed.

Algebra, the language of equations and unknowns,
Unlocks the door to possibilities, all that it owns.
With variables and constants, it weaves its spells,
Solving puzzles and problems, as the story tells.

Through Algebra's lens, patterns emerge,
Logic and reasoning, it does urge.
From quadratic equations to linear graphs,
It teaches us to think, to solve life's tough drafts.

In the realm of science, Algebra shines bright,
It fuels discoveries, like stars in the night.

From physics to engineering, it's a guiding force,
Building bridges, unraveling the laws of the universe.

In finance and economics, Algebra plays its part,
Analyzing data, predicting trends, with mathematical art.
From budgeting to investments, it guides our way,
Making sound decisions, every single day.

So, dear student, embrace Algebra's might,
With perseverance and effort, you'll reach new heights.
Let not the challenges deter your quest,
For Algebra's rewards are truly the best.

Believe in yourself, unleash your potential,
With Algebra as your ally, you'll be influential.
In this journey of learning, you shall surpass,
And conquer Algebra, with brilliance and class.

ELEVEN

FOREVER FREE

In the realm of numbers, where mysteries reside,
Lies the key to unlock the secrets, far and wide.
Algebra, the language of the infinite skies,
Will guide you on a journey, where limits defy.

In science and engineering, its power unfolds,
Equations and formulas, the stories it molds.
From bridges to rockets, it shapes the unknown,
Algebra's magic, in every design, it's shown.

Finance and economics, its relevance is clear,
Analyzing data, forecasting without fear.
Investments and trends, it unravels the path,
Algebra's guidance, a beacon of math.

But it's not just in fields, where numbers align,
Algebra expands the mind, lets imagination shine.

It's a style, a way of thinking, a symphony of thought,
With every problem solved, new horizons are sought.

Embrace the challenges, with perseverance and might,
For Algebra's wonders, will soon come to light.
Believe in yourself, let confidence soar high,
And Algebra's mysteries, you'll surely defy.

So let Algebra be your guide, on this quest you embark,
Unleash the potential, ignite the spark.
Dedicate your effort, with passion and zeal,
And Algebra will show you, what's truly real.

For within its intricate web, lies beauty untold,
A world of possibilities, waiting to unfold.
So fear not, dear student, let Algebra be your key,
And watch as it sets your mind, forever free.

TWELVE

LEAVING NO DREAM BEHIND

In the realm of numbers, where equations reside,
Lies a subject called Algebra, where dreams coincide.
Fear not, dear student, for here I shall proclaim,
The wonders of Algebra, igniting your flame.

Algebra, the key to unlock the unknown,
A language of symbols, in patterns it's sown.
It may seem daunting, with variables unknown,
But unravel its secrets, and you'll reap what's sown.

In the realm of equations, problems to be solved,
Lies the power to conquer, to evolve and evolve.
Algebra, dear student, is not just a chore,
It's a gateway to knowledge, a pathway to more.

For with Algebra, you'll understand the world's design,

From physics and engineering to the stars that align.
From finance and economics to computer science's might,
Algebra is the foundation, guiding us through the night.

So embrace the challenge, let Algebra be your guide,
Unleash your potential, let your brilliance collide.
Believe in yourself, for you hold the key,
To unlock the magic, that Algebra sets free.

Passion and perseverance, your allies on this quest,
Embrace the beauty, and you'll be among the best.
In Algebra lies a universe waiting to unfold,
So let it guide you, as you soar and behold.

For in conquering Algebra, success you shall find,
Unleashing your potential, leaving no dream behind.
So fear not, dear student, for you hold the power,
To conquer Algebra's challenges, and reach the highest tower.

THIRTEEN

CONQUER AND EMBRACE

In the realm of numbers, a journey awaits,
Where Algebra's magic casts its captivating traits.
Fear not, young student, for within lies the key,
To unlock the mysteries that numbers decree.

With equations and variables, a dance will commence,
A symphony of logic, a puzzle to make sense.
Algebra's language, a melody so sweet,
It challenges the mind, a formidable feat.

Through polynomials and functions, a path unfolds,
Where solutions are found, like treasures untold.
In every equation, a problem to solve,
Algebra's embrace, the power to evolve.

From the realms of science to the depths of finance,
Algebra's relevance, a cosmic romance.

It shapes our world, unravels the unknown,
Through numbers and symbols, a wisdom is shown.
    Let Algebra be your guide, your trusted friend,
In its embrace, your spirit will transcend.
Believe in yourself, let perseverance ignite,
For Algebra's beauty will illuminate your sight.
    Embrace the challenges, the trials you face,
For within lies the power to conquer and embrace.
Unlock your potential, let Algebra be your muse,
And witness the magic it has in store for you.
    In realms of numbers, dreams come alive,
Algebra, the foundation on which they thrive.
So, dear student, let this be your call,
Pass Algebra's test, and watch your dreams enthrall.

FOURTEEN

GATHER YOUR COURAGE

In the realm of numbers, where equations dance,
Lies a subject of great power, a glorious chance.
Algebra, dear student, holds secrets untold,
Unlocking the mysteries, make them unfold.

In science's embrace, Algebra takes flight,
Discovering laws that shape day and night.
From atoms to stars, its reach knows no end,
By understanding Algebra, you transcend.

In engineering's domain, Algebra's the key,
Designing wonders, setting imaginations free.
Build bridges, skyscrapers, and machines of might,
With Algebra's guidance, the future shines bright.

In finance and economics, Algebra's might,
Unraveling patterns, revealing trends in sight.

Unlock wealth's door, embrace prosperity's call,
With Algebra's wisdom, you'll conquer it all.

Fear not the unknown, for Algebra's your guide,
Its language, a symphony, where mysteries reside.
With perseverance and grit, you'll conquer each test,
Embrace Algebra's beauty, let it bring you zest.

Believe in yourself, dear student, you'll see,
Algebra's a journey, a chance to be free.
Through challenges faced, you'll grow and evolve,
Unleashing your potential, problems you'll solve.

So gather your courage, with Algebra's embrace,
Forge a path to success, leave no room for space.
Dream big, dear student, let Algebra ignite,
The fire within you, to reach for the height.

For in Algebra's realm, magic and wonder reside,
Open your mind, let it be your guide.
Passion and dedication, your heart shall decree,
Embrace Algebra's power, and set your mind free.

FIFTEEN

BELIEVE IN YOURSELF

In the realm of numbers, where patterns unfold,
Lies the magic of Algebra, a story untold.
A gateway to knowledge, a key to success,
Embrace its challenges, let your doubts suppress.

Algebra, dear student, is a friend so true,
Unleashing your potential, guiding you through.
With determination and a curious mind,
You'll unravel its mysteries, leave no concept behind.

Fear not the equations, the variables unknown,
For within lies a power, waiting to be shown.
Solve the puzzles, unlock the door,
And witness the beauty, like never before.

In science, in art, in the world of finance,
Algebra's presence, a remarkable dance.

It shapes our future, it lights up the way,
Leading us forward, come what may.

    Believe in yourself, my dear student, you'll see,
Algebra's embrace sets your dreams free.
It builds resilience, sharpens your mind,
A tool for success, of the rarest kind.

    So let Algebra be your ally, your guide,
Don't let its challenges cast you aside.
For within its realm, lies a world so vast,
Where dreams are born, and futures are cast.

    Embrace the power, the beauty it holds,
And conquer Algebra, as legends of old.
With every step forward, your brilliance will shine,
A student of Algebra, forever entwined.

SIXTEEN

YOUR VICTORY SONG

In the realm of numbers, where mysteries reside,
Lies a subject called Algebra, a majestic guide.
With its symbols and equations, it may seem complex,
But fear not, dear student, for you shall pass this test.

Algebra, the key to unlock the universe's code,
A pathway to knowledge, where dreams can unfold.
Embrace its challenges, let them fuel your fire,
For within lies the potential to reach higher.

It may seem daunting, with variables unknown,
But with patience and practice, you'll reap what is sown.
Equations and formulas, like puzzles to solve,
Each step you take, your problem-solving evolves.

Believe in yourself, for you hold the power,
To conquer this subject, to bloom like a flower.

With dedication and perseverance, you'll break free,
And witness the magic that Algebra can be.
    Let it be your ally, your guiding light,
Unleash your potential, with all your might.
For Algebra is not just a subject to pass,
It's a tool for success, a window to surpass.
    So, fear not, dear student, for you are strong,
Let Algebra be your melody, your victory song.
Embrace the beauty of numbers, let them dance,
And soon you'll find yourself in the realm of advance.

SEVENTEEN

CONFIDENCE AND GRACE

In the realm of numbers, where magic lies,
There's a subject that unlocks the skies.
Algebra, the key to endless dreams,
Where beauty and potential gleam.

Fear not the equations, my dear student,
For they hold secrets, unseen but prudent.
Embrace the challenges, let them be,
And watch your mind soar, wild and free.

Algebra, a guide through the unknown,
A path to discoveries yet to be shown.
With perseverance and dedication, you'll find,
That solutions to problems, you'll always unwind.

Let Algebra be your ally, your trusted friend,
In its embrace, your doubts will end.

With every equation, a puzzle to solve,
You'll grow wiser and your fears dissolve.

Unlock your potential, let Algebra's might,
Illuminate your path, shining so bright.
For in this enchanted realm, you'll see,
The power and beauty that Algebra holds, key.

So believe in yourself, my student so bold,
Embrace the wonders that Algebra unfolds.
With problem-solving skills, you'll thrive and grow,
And conquer the subject that others may forego.

In the realm of Algebra, your dreams come alive,
A realm where the future, you'll cast and derive.
So take a leap, with confidence and grace,
And let Algebra guide you to a magical place.

EIGHTEEN

ESSENCE OF ALGEBRA

In the realm of numbers, where patterns unfold,
Lies the essence of Algebra, a tale yet untold.
Unlocking the mysteries, a path to explore,
Let your spirit rise, let your knowledge soar.

Algebra, the language of equations profound,
Transforms the abstract into solid ground.
For every problem it presents, a solution awaits,
With perseverance and courage, conquer the gates.

Through variables and constants, equations unite,
Like stars in the sky, shining ever so bright.
In this cosmic dance, you have a role to play,
Embrace the challenge, seize the day.

Algebra, the architect of our modern world,
From bridges to buildings, its wonders unfurled.

With formulas and functions, it shapes our reality,
Let it guide you, embrace its duality.

With every step forward, you build resilience anew,
Algebra is a journey, a chance to renew.
Believe in yourself, let your passion ignite,
And the subject that daunts you, you shall conquer with might.

So, dear student, fear not the unknown,
For Algebra, like magic, shall make itself known.
With dedication and effort, success will be found,
In the depths of Algebra, your potential is crowned.

NINETEEN

FUTURE THAT'S BLESSED

In the realm of numbers, a tale unfolds,
Where Algebra's beauty brightly glows,
A puzzle to solve, a quest to embark,
Let not fear bind thee, let passion spark.

Algebra, a language of patterns and signs,
Unveiling secrets, unlocking the divine,
Equations and formulas, they may seem tough,
But fear not, dear student, for you are enough.

Embrace the unknown, let curiosity guide,
Algebra's dance, let it be your pride,
For within its realm lies infinite might,
The power to conquer, to reach great height.

Through variables and constants, a world takes shape,

In equations and graphs, new pathways escape,
From solving for x to finding unknowns,
Algebra's magic, let it be known.

    Believe in thyself, for you hold the key,
To unlock the wonders that Algebra can be,
With dedication and perseverance, you'll find,
A strength within, a brilliant mind.

    So fear not the numbers, the symbols, the lore,
For Algebra's gates open wide, evermore,
Embrace the challenge, let knowledge unfold,
And watch as your dreams, they begin to behold.

    Passing Algebra, not just a mere test,
But a step towards a future that's blessed,
Believe in thyself, let passion ignite,
And Algebra's beauty, forever take flight.

TWENTY

FIND YOUR WORTH

In the realm of numbers, a world to explore,
Where equations dance, and mysteries restore.
Algebra, the language of the unknown,
Unveiling secrets, a path to be shown.

Fear not the challenges that lie ahead,
For within each problem, knowledge is spread.
Like a puzzle to solve, it tests your might,
But with perseverance, you'll shine so bright.

Oh student, embrace this wondrous art,
Let Algebra ignite the flames in your heart.
For within its depths, lies a hidden key,
Unlocking the gates to endless possibility.

With variables and constants, we shall play,
Simplifying expressions, night and day.

Equations, inequalities, they hold the key,
To unravel the magic, for all to see.
    Algebra, a bridge to the world unknown,
Unleashing potential, like seeds that are sown.
Believe in yourself, let your dreams take flight,
For Algebra's beauty will guide you through the night.
    So study with passion, let your mind unfurl,
For Algebra's wonders, they shall unfurl.
With dedication and effort, you shall prevail,
And conquer the subject, like a triumphant tale.
    Let Algebra be your companion, your guide,
Through the challenges you'll face, side by side.
For within its embrace, you'll find your worth,
And pass this subject, the greatest of Earth.

TWENTY-ONE

POTENTIAL SOAR

In the realm of numbers, let us explore,
A subject that some may find a chore.
Algebra, the key to unlock the unknown,
A journey of equations to call your own.

Oh student, listen to my humble plea,
Embrace the beauty that Algebra can be.
For within its depths lies a world of might,
Where problems are solved and dreams take flight.

Through x and y, and variables unknown,
You'll learn to solve puzzles, like pieces of stone.
Equations and formulas, they may seem tough,
But persevere, my dear, and soon enough,

You'll see the patterns, the logic unfold,
And find the wonders that Algebra holds.

For in this subject, a gift you'll find,
To shape your future and expand your mind.
    From architecture to engineering's core,
Algebra's reach extends, forevermore.
So believe in yourself, and don't despair,
With dedication, you'll conquer the snare.
    Embrace the challenge, let Algebra guide,
As you unravel the secrets it hides.
For in this journey, you'll come to see,
That Algebra is the key to set you free.
    Unlock the wonders, let your potential soar,
With Algebra in hand, success is in store.
So fear not the numbers, let your spirit shine,
For Algebra is a bridge to the divine.

TWENTY-TWO

GREATNESS WILL BE

In the realm of numbers, where mysteries lie,
There's a subject that challenges, reaching for the sky.
Algebra, they call it, a puzzle to unravel,
But fear not, dear student, for you have the power to travel.

Through equations and variables, you'll find your way,
To a world of solutions, where brilliance holds sway.
With dedication as your compass, and perseverance as your guide,
You'll conquer the challenges, standing tall with pride.

Algebra, a language of symbols and codes,
Unlocks the doors to countless roads.
It teaches you logic, and how to think deep,
To analyze, to reason, and secrets to keep.

So embrace the unknown, for within it lies,

A universe of knowledge, where your potential flies.
The numbers may seem daunting, but remember this truth,
In every problem, lies the seed of your youth.

    For Algebra is not just a subject to pass,
But a tool for success, that will forever last.
So rise to the challenge, with passion in your stride,
And let the wonders of Algebra be your guide.

    Believe in yourself, for you have what it takes,
To conquer the unknown, and embrace the mistakes.
With each step you take, closer you'll be,
To unlocking the wonders that Algebra holds for thee.

    So fear not, dear student, for you are strong,
With Algebra as your ally, you'll never go wrong.
For in the realm of numbers, you'll find your key,
To a brighter future, where greatness will be.

TWENTY-THREE

NEVER GIVE IN

In the realm of numbers, a journey begins,
Where logic and beauty dance, as Algebra sings.
Embrace the challenge, let passion ignite,
For within these equations, lies endless insight.

Algebra, the language of the universe's code,
Unveiling secrets, ready to be bestowed.
Through variables and constants, patterns unfold,
Unlocking the mysteries, as our story is told.

With determination as your guiding light,
Conquer the equations, with all your might.
For in the depths of struggle, growth is found,
A metamorphosis, where brilliance is crowned.

Believe in yourself, for you hold the key,
To unravel the truth, and set your mind free.

From quadratic equations to polynomials grand,
Discover the power that lies in your hand.
    Algebra, the bridge to a future untold,
Where dreams become real, and stories unfold.
A gateway to success, it paves the way,
To a world of opportunities, come what may.
    So fear not the numbers, let them be your guide,
For Algebra's beauty, you can't hide.
Embrace its challenges, with open arms,
And watch as your potential brightly warms.
    For Algebra is not just a subject to pass,
But a journey of growth, where you'll amass,
The skills and knowledge to conquer the unknown,
And stand tall, with victories to be shown.
    So don't falter, dear student, take this chance,
To dive into Algebra's wondrous expanse.
Believe in yourself, and never give in,
For Algebra's treasures, it's time to begin.

TWENTY-FOUR

DON'T LET DOUBTS PERSIST

In the realm of numbers, where patterns reside,
Lies the enchanting world of Algebra, side by side.
Fear not this subject, for it holds a key,
To unlock the doors of possibility.

With variables and equations intertwined,
Algebra's beauty is not hard to find.
It's a language of logic, a puzzle to solve,
A challenge to conquer, a mystery to evolve.

Within its depths, secrets are concealed,
Solutions waiting to be revealed.
Embrace the unknown, let your mind explore,
Algebra's treasures, forevermore.

Fear not the x's and y's that may confound,
For in their embrace, knowledge is found.

Each problem you solve, a victory to claim,
A step closer to success, acclaim.
    So, dear student, don't let doubts persist,
For Algebra's power, you can't resist.
Believe in yourself, let your potential shine,
Pass this subject, and the world will be thine.
    With Algebra as your ally, you'll soar high,
Able to reach the stars in the sky.
So, take this chance, dive into the fray,
Unlock the wonders of Algebra today.

TWENTY-FIVE

JOURNEY OF DISCOVERY

In the realm of numbers, where mysteries reside,
There lies a subject, often feared and denied.
Algebra, they call it, a puzzle to unfold,
With its unknown variables, stories yet untold.

Fear not, dear student, for within these walls,
Lies a world of wonders, where potential calls.
Algebra, oh Algebra, the beauty it holds,
Unlocking paths untrodden, where dreams unfold.

Equations and formulas, like a symphony they dance,
Melodies of logic, giving life a chance.
Solving for x, a quest for the unknown,
A journey of discovery, all on your own.

Algebra, a language, a code to decipher,

A tool for problem-solving, your mind's enhancer.
Through its trials and challenges, you will grow,
Building resilience, a strength you'll come to know.

Embrace the challenge, with courage in your heart,
For Algebra's treasures, they're waiting to impart.
Believe in yourself, for you hold the key,
To unlock the greatness that lies within thee.

So, fear not the numbers, the symbols, or the signs,
For Algebra's embrace, it's a gift divine.
Pass this subject, and you'll pass through the door,
To a world of endless possibilities, forevermore.

In Algebra's realm, success awaits,
A brighter future, beyond the classroom gates.
Embrace the challenge, let your spirit soar,
For Algebra's triumphs, they'll be yours to explore.

TWENTY-SIX

FEAR NOT THE X'S OR THE Y'S

In the realm of numbers and equations,
Where Algebra weaves its mystical sensations,
Lies a path to unlock your potential and might,
A journey that will fill your future with light.

Algebra, the language of the unknown,
A code to decipher, a seed to be sown,
It may seem daunting, with symbols and signs,
But fear not, dear student, for greatness it aligns.

Through variables and formulas, you'll find,
A world of solutions, a brilliant mind,
From quadratic equations to polynomials grand,
A universe of possibilities at your command.

Algebra, the key to unravel the unknown,
A tool to conquer challenges, to call your own,

With every problem solved, your confidence will grow,
And the seeds of success, within you, will sow.
    So let not frustration or doubt cloud your way,
Embrace the numbers, the puzzles they convey,
For Algebra holds the secrets to success,
A gateway to triumph, a path you must address.
    Believe in yourself, let your spirit rise,
Embrace the challenge, with determined eyes,
For within the realm of Algebra's art,
Lies the power to shape your very own chart.
    So fear not the x's or the y's,
For Algebra is where your potential lies,
Pass this subject, and you'll truly see,
The wonders and joys of mathematical glee.

TWENTY-SEVEN

BRIGHT AND TRUE

In the realm of numbers, where equations reside,
Lies the gateway to knowledge, where dreams coincide.
Algebra, dear student, is not just a chore,
But a powerful tool, your mind can explore.

Through variables and constants, a language is revealed,
A world of patterns and logic, where solutions are concealed.
Embrace the challenge, let doubts fade away,
For Algebra holds treasures, you'll discover one day.

Beyond the classroom, in the real world outside,
Algebra unlocks the doors, where success will abide.
From engineering marvels to financial might,
Algebra paves the way, to reach new heights.

So when the equations seem complex, and thoughts become unclear,
Remember, dear student, you're stronger than you appear.
With perseverance and dedication, you'll conquer each test,
And Algebra's mysteries, you'll lay to rest.

Believe in your potential, let your brilliance shine,
For Algebra is a journey, your spirit will define.
Embrace the unknown, let curiosity guide,
And Algebra's secrets, you'll uncover with pride.

So fear not, dear student, for you hold the key,
To master the language, and set your mind free.
Passing Algebra is not just a goal to pursue,
It's a stepping stone to a future that's bright and true.

TWENTY-EIGHT

PASSION AND DRIVE

In the realm of numbers, where mysteries unfold,
Lies the power of Algebra, waiting to be told.
Fear not, young student, for I shall impart,
A tale of triumph, a song from my heart.

Algebra, a puzzle of symbols and signs,
Unleashes the mind, where brilliance aligns.
It's not just about numbers, my dear friend,
It's a language of logic, a means to transcend.

Unlock the secrets, embrace the unknown,
For Algebra's beauty shall soon be shown.
Through equations and formulas, you shall see,
The world's hidden patterns, so clear and free.

From solving for x to graphing a line,
Algebra's journey will be truly divine.

It teaches perseverance, resilience, and might,
To conquer your doubts and shine so bright.
  With Algebra as your guide, you'll soar high,
Defying the limits, reaching for the sky.
In every problem lies a chance to grow,
To cultivate knowledge and let your spirit flow.
  So, dear student, with passion and drive,
Embrace Algebra's wonders, let your dreams thrive.
For in this subject, lies the key,
To unlock your potential, and set your mind free.

TWENTY-NINE

STEPPING STONE

In the realm of Algebra, where numbers dance,
Lies a treasure trove, a captivating chance.
Fear not the unknown, for within its embrace,
Lies the power to solve life's intricate maze.
    Algebra, a language of patterns and codes,
Unveiling secrets, as the story unfolds.
It may seem daunting, with its symbols and signs,
But fear not, dear student, for greatness it defines.
    Embrace the challenge, let curiosity lead,
Unlock the wonders, let your brilliance succeed.
For within the equations, lies a world unseen,
Where logic and reason create a vibrant scene.
    Algebra is a key, to unlock your mind,
To solve problems, and new solutions find.

It shapes the future, molds the world around,
With its power, infinite possibilities abound.

    So persevere, dear student, let your doubts fade,
With every step forward, the path is laid.
Believe in yourself, for you hold the key,
To conquer Algebra, and set your spirit free.

    Let not fear or doubt hinder your way,
Embrace the challenge, let your brilliance sway.
For passing Algebra is not just a goal,
But a stepping stone to a future untold.

    So, my dear student, don't shy away,
Embrace Algebra's journey, let your brilliance sway.
For within its embrace, lies your potential untold,
Unlock the treasures, let your brilliance unfold.

THIRTY

YOU WILL RISE

In the realm of numbers, where patterns reside,
Lies a subject feared, but let not your spirits hide.
Algebra, the key to unlock the unknown,
A world of possibilities, yet to be shown.

    Embrace the challenge, let doubts fade away,
For in this pursuit, greatness lies in sway.
Equations and variables may seem a haze,
But within lies the power to amaze.

    Like a puzzle, Algebra unveils its grace,
A language of symbols, a dance to embrace.
With patience and persistence, you'll find the way,
To conquer the obstacles, day by day.

    Fear not the x's and y's that may confound,
For they hold the secrets waiting to be found.

Solve the equations, unravel each knot,
And discover the beauty that Algebra has got.

Passing this subject is more than a test,
It's a stepping stone to a future, the best.
With Algebra's might, you'll shape the world,
Unleashing your brilliance, like a flag unfurled.

So let your potential shine, like a guiding star,
Harness the power that Algebra can spar.
For within lies the strength to find new solutions,
To unravel mysteries and endless revelations.

Believe in yourself, and let doubts be no more,
Unlock the wonders of Algebra's core.
With resilience and perseverance, you will rise,
And conquer the subject, to endless skies.

THIRTY-ONE

DOUBTS WILL SURELY BREAK

In the realm of numbers, where equations reside,
There lies a world of wonder, where dreams collide.
Algebra, they call it, a puzzle to unravel,
A gateway to knowledge, a path we must travel.

At first, it may seem daunting, a maze of unknown,
But fear not, dear student, for you are not alone.
With every problem faced, a lesson is learned,
A chance to grow and flourish, your skills will be earned.

Like a painter with a brush, you'll craft solutions anew,
Unveiling hidden patterns, revealing what is true.
Through variables and constants, you'll navigate the way,

Unlocking doors of understanding, each and every day.

Algebra is a language, a symphony of signs,
Where x's and y's dance, intertwining their lines.
Embrace the challenge, let curiosity ignite,
For in the depths of algebra, you'll find your inner light.

So, persevere, dear student, with every step you take,
You'll conquer the unknown, your doubts will surely break.
Believe in your abilities, let passion be your guide,
And witness the magic that Algebra will provide.

For within its realm, lies beauty untold,
A world of endless possibilities, waiting to unfold.
So, let not fear nor doubt hold you back,
Embrace the journey, let Algebra be your track.

Passion and persistence, the keys to your success,
With them, dear student, you'll surely progress.
So, hold your head up high, let your spirit soar,
For you have the power to conquer Algebra's lore.

THIRTY-TWO

DEAR STUDENT

In the realm of numbers, there lies a grand key,
Unlocking the wonders of Algebra, you see.
With symbols and equations, it may seem tough,
But trust in your brilliance, you'll rise high enough.

Algebra, dear student, holds secrets untold,
A language of patterns, a treasure to behold.
It's a bridge to the future, a path to success,
Embrace its challenges, and you'll surely impress.

In every equation, a puzzle to solve,
A chance to evolve, to let your mind revolve.
With variables and constants, you'll navigate,
And conquer each problem, no matter how great.

For in Algebra's realm, the power awaits,
To shape the world with your mathematical traits.

From equations to graphs, you'll find endless ways,
To innovate, create, and set your dreams ablaze.

So, fear not the unknown, embrace the unknown,
With perseverance and courage, you'll be shown.
That Algebra's beauty lies deep within you,
Unleash your potential, let your brilliance shine through.

Believe in yourself, in your infinite might,
With Algebra as your guide, you'll reach new heights.
Pass this subject, and you'll open the door,
To a future where possibilities will soar.

So, dear student, keep pushing, don't ever give in,
For Algebra's magic lies under your skin.
Let your passion ignite, let your spirit take flight,
And conquer this subject with all of your might.

THIRTY-THREE

ALGEBRA DIVINE

In the realm of numbers, where equations dance,
Lies the power to unlock a vast expanse.
Algebra, a language both elegant and true,
A gateway to dreams waiting for you.
    With variables and constants, you'll navigate,
A world of patterns, where wonders await.
Fear not the x's, the y's, or the z's,
For Algebra's embrace will set your mind at ease.
    When equations unravel their intricate seams,
Magic unfolds within your wildest dreams.
Like a sculptor, you'll mold solutions with care,
Crafting beauty from numbers, beyond compare.
    The unknowns may challenge, but don't despair,
For Algebra's secrets, you'll learn to declare.

With perseverance and a heart full of might,
The answers will shimmer, revealing their light.
    So, let not frustration cloud your sight,
Embrace the challenge, let your brilliance ignite.
For Algebra's power is not just a test,
It's a stepping stone to a future, the very best.
    Through Algebra's prism, you'll find your place,
A world of innovation, yours to embrace.
Believe in yourself, let passion be your guide,
And witness the magic Algebra will provide.
    With courage and perseverance, you shall soar,
Unleashing treasures you've never seen before.
Unlock your potential, let your brilliance shine,
And bask in the glory of Algebra divine.
    For in this subject lies a world untold,
Where possibilities bloom and dreams unfold.
Passing Algebra is not just a goal,
It's the key to a future that's yours to mold.

THIRTY-FOUR

SYMBOLS AND SIGNS

In the realm of numbers and equations,
Where logic and patterns dance,
Lies the realm of Algebra,
A subject that holds a powerful stance.

Fear not the unknown variables,
For within them lies a hidden key,
Unlocking doors to boundless wonders,
Where possibilities are set free.

Algebra, a language of the universe,
A symphony of symbols and signs,
It weaves a tapestry of understanding,
Where brilliance and intellect align.

With every step, you grow stronger,
As you conquer each daunting test,

Embrace the challenges that come your way,
And let your determination manifest.
    For in the realm of Algebra's embrace,
Lies a world of innovation and creation,
Where minds are shaped and dreams take flight,
A journey of endless exploration.
    So let your brilliance shine through,
As you unravel each mathematical maze,
Passing Algebra is not just a goal,
But a stepping stone to brighter days.
    Believe in your abilities, my friend,
And embrace the treasures Algebra holds,
For with perseverance and hard work,
Success and opportunities unfold.

THIRTY-FIVE

YOUR FAITHFUL FRIEND

In the realm of numbers, where secrets reside,
Lies the enchantment of Algebra, waiting inside.
Though equations may seem like a daunting sea,
Unlock their power, and set your spirit free.

    Algebra, dear student, is not just a chore,
But a language that opens infinite doors.
It's a symphony of symbols, a puzzle to solve,
A journey of discovery, where mysteries evolve.

    Embrace the unknown, let your fears be erased,
For Algebra's embrace is a sacred space.
It teaches you logic, sharpens your mind,
Equipping you with skills, rare and refined.

    Through x's and y's, equations unfold,
Revealing patterns and stories untold.

Solve for the variables, find the unknown,
And witness the marvels that will be shown.
   Algebra, dear student, is your secret key,
To unlock a world of possibility.
It's the bridge to physics, the language of science,
A force that propels dreams into reliance.
   So fear not the numbers, let them inspire,
For Algebra's beauty will never expire.
With diligence and passion, you'll surely amass,
The knowledge and wisdom that this subject en masse.
   Believe in yourself, for you have the might,
To conquer Algebra and shine oh-so-bright.
Let this poem be your guide, your faithful friend,
And may your journey through Algebra never end.

ABOUT THE AUTHOR

Walter the Educator is one of the pseudonyms for Walter Anderson. Formally educated in Chemistry, Business, and Education, he is an educator, an author, a diverse entrepreneur, and he is the son of a disabled war veteran. "Walter the Educator" shares his time between educating and creating. He holds interests and owns several creative projects that entertain, enlighten, enhance, and educate, hoping to inspire and motivate you.

> Follow, find new works, and stay up to date
> with Walter the Educator™
> at WaltertheEducator.com

www.ingramcontent.com/pod-product-compliance
Lightning Source LLC
LaVergne TN
LVHW010602070526
838199LV00063BA/5053